I0815272

WILD STEM

STEM in the OCEAN

Megan Borgert-Spaniol

Checkerboard Library

An Imprint of Abdo Publishing
abdobooks.com

abdobooks.com

Published by Abdo Publishing, a division of ABDO, PO Box 398166, Minneapolis, Minnesota 55439.

Printed in the United States of America, North Mankato, Minnesota
102023
012024

Design: Denise Hamernik, Mighty Media, Inc.
Production: Mighty Media, Inc.
Editor: Anna Anderhagen
Cover Photographs: Max Topchii/Shutterstock Images
Interior Photographs: Andrea Izzotti/Shutterstock Images, p. 17; fenkieandreas/Shutterstock Images, p. 27; Luis Javier Sandoval/VWPics/AP Images, p. 23; MaeManee/Shutterstock Images, p. 5; Mighty Media, Inc., pp. 10–11, 20–21; NOAA Office of Ocean Exploration and Research/AP Images, p. 15; P Songthongweerakul/Shutterstock Images, p. 7; panaramka/iStockphoto, p. 25; Ralph White/Getty Images, p. 13; Richard Whitcombe/Shutterstock Images, p. 29; Stanislav Moroz/Adobe Stock, p. 9; Yvonne Werner/Shutterstock Images, p. 19
Design Elements: Lorthois Yuliya/Shutterstock Images; supanut/Adobe Stock

Library of Congress Control Number: 2023939336

Publisher's Cataloging-in-Publication Data
Names: Borgert-Spaniol, Megan, author.
Title: STEM in the ocean / by Megan Borgert-Spaniol
Description: Minneapolis, Minnesota : Abdo Publishing, 2024 | Series: Wild STEM | Includes online resources and index.
Identifiers: ISBN 9781098291976 (lib. bdg.) | ISBN 9781098278878 (ebook)
Subjects: LCSH: Science--Study and teaching--Juvenile literature. | Technology--Study and teaching--Juvenile literature. | Engineering--Study and teaching--Juvenile literature. | Mathematics--Study and teaching--Juvenile literature. | Ocean--Juvenile literature.
Classification: DDC 577.0--dc23

CONTENTS

WILD OCEAN

Do you love the sound of waves and the smell of salt? Do you dream of exploring the vast depths of the sea? If so, you might enjoy a career that takes you to the ocean!

Many scientists work in the ocean. Oceanographers study the ocean from its surface to the seafloor. Marine archaeologists uncover the past through shipwrecks and underwater **artifacts**. Marine biologists study ocean creatures, while marine veterinarians care for these animals. Marine conservationists protect the ocean and all its **inhabitants**.

What kind of scientist would you like to be? There's only one way to find out. Zip up your wet suit and grab a snorkel. It's time to explore science, technology, engineering, and math (STEM) in the ocean!

STEM
Science
Technology
Engineering
Math

Did you know that humans have explored only 5 percent of the ocean?

OCEANOGRAPHERS

Many scientists refer to the ocean as "inner space." Like outer space, there is so much to explore! Scientists who study this vast inner space are called oceanographers.

Geological oceanographers study the ocean floor and its structures. These structures include mountains, volcanoes, and other undersea features. Geological oceanographers may also study rock samples from the seafloor. These rock samples help scientists understand how Earth has changed over millions of years!

Chemical oceanographers study the ocean's chemistry and how it affects marine life. For example, the ocean absorbs carbon dioxide from the atmosphere. Over time, excess carbon dioxide makes the ocean more acidic. Scientists examine how ocean acidity affects coral reefs and other sea life.

Losin Island is an underwater mountain. Only the very top sticks out above the water. It is home to deep coral reefs and various fish.

Physical oceanographers study temperatures, tides, and currents in the ocean. They also learn how ocean temperatures and currents drive global weather. For example, some warmer ocean currents fuel hurricanes. The more scientists understand these warmer currents, the faster they can warn people to take shelter.

Many oceanographers work on research ships to gather data about the ocean. These scientists use various tools to conduct research. Oceanographers use remote-controlled vehicles to collect samples of rocks and water. **Satellites** help oceanographers map the **topography** of the ocean floor. With time and technology, oceanographers will slowly reveal the secrets of inner space!

Natural events, such as underwater volcano eruptions, cause changes in ocean water. Oceanographers collect water samples and study how sea life adapts to these changes.

GET WILD!

OCEAN CURRENTS

In this experiment, watch how hot and cold water interact to understand what drives deep ocean currents. Cold water is more **dense** than warm water. As the cold water sinks, warmer water moves in to take its place. This is how ocean currents circulate water around the world!

WHAT YOU NEED:

- ✓ large clear container
- ✓ water
- ✓ 2 paper cups
- ✓ ice cubes
- ✓ blue and red food coloring
- ✓ microwave

WHAT YOU DO:

1. Fill the container about halfway with room temperature water.
2. Place ice cubes into a paper cup. Squeeze four drops of blue food coloring over the ice cubes.
3. Fill the other cup with water. Heat it in the microwave for one minute. Be careful when handling the hot cup. Add four drops of red food coloring.
4. Pour the ice cubes into one end of the large container. The blue food coloring will show you the movement of the cold water.
5. Slowly pour the hot water into the opposite end of the large container. The red food coloring will show you the movement of the hot water.
6. Observe how the cold water moves with the hot water through the container. What do you notice?

MARINE ARCHAEOLOGISTS

In 1985, an underwater camera cruised above the floor of the Atlantic Ocean. It passed over something massive. It was the RMS *Titanic*, a passenger ship that had sunk more than 70 years earlier! Since this discovery, marine archaeologists have recovered thousands of **artifacts** from the RMS *Titanic* shipwreck. Many are personal items, such as shoes, jewelry, and clothing.

Shipwrecks like the RMS *Titanic* provide a wealth of information to marine archaeologists. These scientists study human history through objects found in the ocean. They research underwater sites holding shipwrecks, sunken airplanes, fishing structures, and other artifacts lost at sea.

The RMS *Titanic* was found at about 12,500 feet (3,810 m) below sea level. Can you find the spare anchor on the RMS *Titanic*?

Marine archaeologists use various technologies to study underwater sites. For example, multibeam sonar uses sound waves to map the seafloor and any objects on it. Magnetometers detect magnetic materials, such as anchors made with iron.

Marine archaeologists can scuba dive as far as 130 feet (40 m) deep. For deeper ocean sites, scientists use remotely operated vehicles (ROVs). These vehicles collect photos and videos of a site. Then, scientists combine these images to re-create the underwater site digitally. This digital creation preserves the site as an undersea museum that gives humans a glimpse of the past.

This ROV is named *Deep Discoverer*. It can dive 3.7 miles (6 km) deep!

MARINE BIOLOGISTS

The ocean is full of life, from colorful corals to tentacled squids. Scientists who study sea life are called marine biologists. Many marine biologists focus their research on a type of organism, such as jellyfish. Or they might focus on a specific species, such as the lion's mane jellyfish.

If you could study any ocean creature, which would you choose? Whales or dolphins? Seals or walruses? Mammals like these are popular subjects in marine biology. But scientists must also pay attention to much smaller ocean creatures. Some marine biologists study organisms they can't even see without a microscope!

Dolphins are social animals. They communicate with whistles, clicks, squeaks, barks, groans, and yelps.

Whatever size creature they study, marine biologists need tools to conduct research. Some scientists scuba dive with waterproof cameras. In deeper waters, researchers use ROVs to gather images. Some ROVs have armlike parts that can pick up objects and collect samples.

Marine biologists use tags to study larger marine species, such as whales and sharks. Researchers attach a tag to an animal using clips, straps, or strong glue. The tag creates signals picked up by radio or **satellite**. Tags allow scientists to track animals' movements. Tags can also record an animal's heart rate and vocalizations.

In 2019, Australian researchers collected humpback whale snot by flying a drone above the whale's blowhole!

GET WILD!

BLUBBER

Whales have a thick layer of fat under their skin called blubber. Feel the insulating effects of blubber with this experiment!

WHAT YOU NEED:

- ✔ large bowl
- ✔ cold water
- ✔ ice cubes
- ✔ 4 plastic bags
- ✔ spatula
- ✔ vegetable shortening
- ✔ 2 rubber bands

WHAT YOU DO:

1. Fill the bowl about halfway with cold water and ice cubes.
2. Place one of your hands inside a plastic bag. Use a spatula to coat the outside of the bag with shortening.
3. Place a second plastic bag over the first. Secure them around your wrist with a rubber band.
4. Have a helper place two plastic bags over your other hand and secure them around your wrist with a rubber band.
5. Place both your hands into the bowl of ice water. Which hand is warmer? How long can you hold each hand in the water?

MARINE VETERINARIANS

Like land creatures, ocean animals experience illness and injury. Scientists who provide health care to ocean animals are called marine veterinarians. They examine, **diagnose**, and treat their aquatic patients. Marine vets **specialize** in a type of animal, such as mammals, fish, or **invertebrates**.

Marine veterinarians spend a lot of time in the water. They must be skilled at swimming, diving, snorkeling, and operating watercraft. Sometimes, they care for animals in the wild. They might lift an octopus onto their boat to draw a blood sample. They also might bring an animal to a medical facility for X-rays or surgery. Whether healing an injured penguin or **monitoring** a dolphin's birth, marine vets help ocean life thrive!

Marine veterinarians measure a juvenile shark. After taking blood and tissue samples, they will release the shark back into the ocean.

MARINE CONSERVATIONISTS

Life on Earth depends on a healthy ocean. The sea is a source of oxygen and food. While humans need the ocean, many human activities threaten it.

One of the biggest threats to the ocean is climate change. When humans burn fossil fuels for energy, greenhouse gases are released into the atmosphere. These gases trap heat on Earth's surface. This heat increases ocean temperatures, affecting marine species. Other human-caused threats to the ocean include oil spills, plastic pollution, and overfishing.

Not all human actions harm the ocean. Scientists who work to protect and restore ocean **habitats** are called marine conservationists. These scientists learn about the ocean and marine life to protect them.

In our oceans, there are huge floating plastic islands. The Great Pacific Garbage Patch has 1.8 trillion pieces of floating plastic!

Marine conservationists perform a wide range of activities to protect the ocean. Some marine conservationists research ways to clean up oil spills. Others develop methods of removing plastic from the ocean. Still other marine conservationists restore coral reefs.

Marine conservationists use various tools to conduct ocean research. Underwater flashlights aid scientists in studying tiny corals. Solar-powered drones collect ocean data and detect oil spills. **Infrared** cameras help scientists track whales at night. With the help of technology, marine conservationists can better understand and **advocate** for our amazing oceans.

Some marine conservationists grow corals in protected areas and then transplant them onto reefs.

A CAREER IN THE OCEAN

Do you have a passion for STEM in the ocean? There are many career paths to choose from. Each path leads to a wide range of opportunities.

A marine veterinarian might work at an aquarium, research center, or ocean **sanctuary**. A marine archaeologist could work for a museum or university. An oceanographer might work for the government or a weather service. No matter where they work, these scientists share a love of our planet's largest ecosystem.

Maybe one day, the ocean will be your office!
What will you discover?

GLOSSARY

advocate—to defend or support a cause.

artifact—an object remaining from a particular location or time period.

dense—having a high mass per unit of volume.

diagnose—to recognize something, such as a disease, by signs, symptoms, or tests.

habitat—a place where a living thing is naturally found.

infrared—energy transmitted by waves, which can be felt as heat.

inhabitant—one who lives permanently in a place.

invertebrate—an animal without a backbone.

monitor—to watch, keep track of, or oversee.

sanctuary—a refuge for wildlife where predators are controlled and hunting is illegal.

satellite—a manufactured object that orbits Earth. It relays scientific information back to Earth.

specialize—to concentrate one's efforts in a specific area.

tentacled—describing an animal with long, flexible structures called tentacles. Tentacles are used for feeling or grasping.

topography—the shape, height, and depth of the features of a place. A topographic map indicates these features.

ONLINE RESOURCES

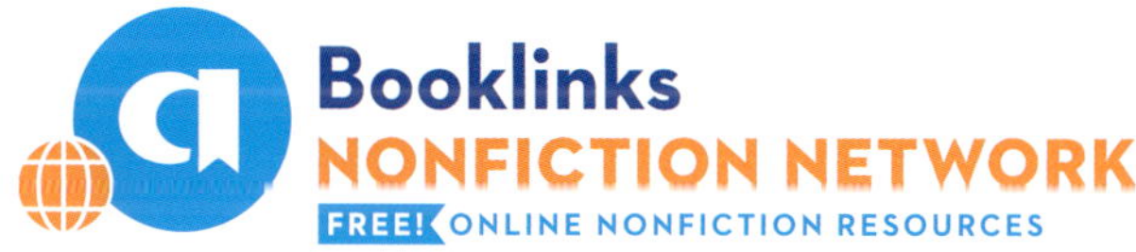

To learn more about STEM in the ocean, visit **abdobooklinks.com**. These links are routinely monitored and updated to provide the most current information available.

INDEX